HERBS AND FRUIT FOR DIETING

Presents a selection of herbs which can reduce weight and, if taken with various fruits, can actually prevent the formation of superfluous fat.

HERBS AND FRUIT
FOR DIETING

by

CERES

Drawings by Alison Ross

1981 SHAMBHALA *Boulder*

SHAMBHALA PUBLICATIONS, INC.
1920 Thirteenth Street
Boulder, Colorado 80302

Published by arrangement with Thorson Publishers, Ltd.
Distributed by Random House

Printed in the United States of America

Library of Congress Cataloging in Publication Data
Ceres.
  Herbs and fruit for dieting.

  (Everybody's home herbal)
  Originally published under title: Herbs and fruit
for slimmers.
  Includes index.
    1. Reducing diets. 2. Herbs—Therapeutic use.
3. Fruit—Therapeutic use. 4. Medicinal plants.
I. Title.
RM222.2.C36   1981   616.3'98061   80-53452
ISBN 0-87773-204-3 (pbk.)
ISBN 0-394-74837-9 (Random House: pbk.)

CONTENTS

INTRODUCTION

By now everyone knows that serious slimming means eating less. But what is not generally known, it seems, is that the quality of the food consumed can make up for the quantity.

It is no good pretending that any dieting is easy. Crash diets, especially if friends are on them too, can be quite amusing for a few days, but too often they only consist of a weight-losing and weakening variety of foods which are not at all satisfying and are therefore quickly given up.

It is much better to try to lose weight slowly and regularly while living on a health-giving and high quality diet which includes the necessary balance of nutritious ingredients. The most difficult part, at the beginning, is to cut out sugar and other carbohydrates and animal fats, although substitutes in the form of fresh and dried fruits and vegetable oils become quickly just as pleasing.

The most important thing, if you know you are too fat and fleshy, or that you have too much 'adipose tissue' as the Victorians so delicately said, is to find out the reason why your body is carrying round the extra burden. It is essential, particularly if you are not feeling fit, that you ask your doctor or a qualified medical herbalist for their advice before you start on any slimming campaign.

They may tell you that your whole system is sluggish and that you are retaining liquids because your metabolism is slowed up or there may be a lack of glandular balance or malfunctioning of other organs.

You may be eating the wrong foods altogether and far too much of them as well. You may easily be deficient in vitamins or mineral salts and herbs and fruit will certainly provide these.

You may have to face the truth that you are under-exercised and the remedy for that, provided you are not physically handicapped, lies entirely in your own hands, or feet.

But to go back to herbs and fruit. This book tells how some herbs can help to get rid of superfluous flesh and how, if they are taken together with the various fruits of the earth like Grapefruit, Avocado Pears, Tomatoes, Peppers and Mushrooms, can even help to prevent the formation of unwanted layers of fat. It will remind you that raw fruit and vegetables have twice the vitamin content of cooked ones.

Slimming, or anti-fat, herbs can be taken in the pleasant form of teas or tisanes and they work in a variety of ways. There is a note at the end of this Introduction describing how these can be made.

The herbs work differently. Some are tonics; some digestives; some stimulate the glands; others speed up the metabolism without, as drugs might, doing it any harm. There are herbs that are mild diuretics which help the kidneys to get rid of impurities and liquids that might hang about the system too long and 'blow it up'.

There are many beneficial flavouring herbs that can help to make a slimming diet more interesting to the taste and there are even herb seeds and stems and fruits to chew in moments of panic when the desire for a sweet or a snack is desperate. Some people carry a 'snack bag' containing Angelica or Celery stems, or Dill or Fennel seeds, which taste of aniseed, and a few Sunflower seeds, to stop themselves from rushing for a sweet, biscuit, bun or bit of cake.

Some of our commonest plants, some people even call them weeds, like Dandelions and Cleavers, are the most useful herbs of all. The word 'herb', by the way, is taken in this series of books to mean the buds, flowers, seeds, fruits, leaves, stems and roots of any traditionally tested and approved beneficial remedial or culinary plant.

But as well as drinking herbal teas, one of which, Maté Tea, is recommended for almost everyone because it has the reputation of being so filling and satisfying and 'good for the intellect', all would-be slimmers should have one fruit meal and another of raw vegetables and salad plants every day. The variety of the latter, with the addition of protein-providing lean meat, fish or poultry, cheese or eggs and a little vegetable oil, can be interesting, palatable and satisfying, both physically in bulk, and mentally in texture and flavour.

It is most important, especially for slimming weight-watchers, to eat *slowly* and to chew all food well so that the digestive juices can get started before the food reaches the stomach.

It is also very important not to rush at a new diet but to give the system time and a chance to adjust by starting it gradually. Many people have found it best to start with an all-fruit breakfast; you will be surprised at how pleasant and satisfying an apple or two, some grapes or a pear (and do try to eat their skins after they have been washed), plus a banana, can be. This fruit breakfast lasts just as long, indeed far longer, than some toast and a cup of ordinary coffee. Try a cup of Dandelion coffee, so much better for you than the 'Arabian berry' at mid-morning. Don't expect it to taste exactly like the coffee you are used to, for it has its own flavour which you will come to like. If you must sweeten it, add a small spoonful of honey instead of sugar.

At first, after your fruit breakfast, start cutting down on all sweetened foods and on potatoes and bread until you have cut them out altogether and have replaced them with sweet fruits, fresh mixed salads and raw vegetables. Eat crispbreads instead of ordinary bread, but do not have too much, and if you hanker after some form of cereal, try muesli together with wheatgerm, which provides vitamin B.

Do not try to cut down on your usual protein intake. Indeed, you may find at first that you want to eat more cheese and eggs and meat, but as you get used to your new way of feeding, which after all is only cutting out the harmful, fat-forming foods, you will be amazed at how quickly you realize that you are feeling far better and more alert and able to cope with the stress and strain of modern life.

Don't weigh yourself very frequently! Wait a month and then give yourself a surprise by standing on the scales first thing in the morning, before your fruit breakfast, and then perhaps weigh yourself every week, and in six months, or a year, when your weight has steadied, look at an old photograph of yourself and get a glow of righteous self-satisfaction at how much younger, healthier and slimmer you look now.

How you have, in fact, substituted fitness for fatness, and you will wonder why you did not do it years ago!

The herbs mentioned in the text are usually supplied by qualified medical herbalists, but can also be obtained, on their recommendation, from health food stores or herbal shops. Unless otherwise stated on the packet, the usual way of making a herbal tea, tisane or infusion, is to pour one pint of boiling water on to one teaspoonful of the fresh or dried herb.

If you find it essential to sweeten your drink, always use a little honey, not sugar and NEVER artificial sweeteners.

1
ANGELICA
(Angelica archangelica)

This tall, large-leaved herb, with its great convex umbels of small greenish-white flowers, is easy to grow but is ground-hungry. It needs constant replanting in different places.

Although the bright green stems may be very familiar as a cake decoration when they have been candied in unlimited sugar, Angelica is useful as a herb for slimmers in several less fattening, sugarless ways. Its young leaves can be eaten in salads; and a tea can also be made from them and from the young stems, which acts as a general tonic and stimulator. It is good for helping the digestion after a bulky salad, as an old herbal says: 'It is grateful to the stomach.'

Angelica tea also has the curious reputation of helping to produce a disgust for liquor in heavy drinkers.

The seeds from this plant can be chewed instead of sweets, like those of Fennel (page 36), or Dill (page 36), Parsley (page 49) and Sunflowers (page 58).

A small amount of chopped young Angelica stem in a omelet or mixed with cottage cheese provides a pleasant, unusual flavouring.

The name 'Angelica' is supposed to have been started because of the tradition that an angel first revealed the plant's antiseptic powers and properties during a plague epidemic: 'Uppon what occasion this excellent name was first given unto it for the excellent vertues thereof, and for that God made it knowne to men, by the ministrie of an angel', wrote Thomas Brasbridge in 1578, in his *Poore Man's Jewel.*

Gerard, a more famous sixteenth-century herbalist, said:

Angelica is a singular remedy against poyson ... if you doe but take a piece of the root and hold it in your mouth, or chew the same between your teeth, it doth most certainly drive away the pestilential aire, yea, although the corrupt aire have already possessed the hart, yet it driveth it out again.

Parkinson, in his *Paradisus in Sole*, put it first among the 'Herbes of the most especiall use that a Country Gentlewoman should grow'. In his day a syrup was made by gashing the hollow green stems, filling them up with sugar and cutting them down in three days' time, thus making a 'most delicate [but fattening] confection'.

Angelica can be juiced, without sugar, to add and improve the flavour of other vegetable juices or soups.

Slimmers, of course, should give up all thick soups and only take clear soups, consommés or broths.

Angelica
(Angelica archangelia)

2

AVOCADO PEAR
(Persea gratissima)

Avocado Pears belong to the same botanical family as the Bay tree and came originally from tropical South America. The Aztecs used them for food and called them 'Aguacates' or 'Alligator Pears'.

They were gradually introduced to other hot countries and even grown in Britain as early as 1739, in glasshouses, as 'stove evergreen shrubs'. No one seems to have recorded whether they ever fruited in these conditions. Nowadays we import them from various places including South Africa (in some parts there they grow in gardens as commonly as apple trees in Britain), and from Israel.

There are several forms and varieties cultivated and the shape of Avocadoes can vary from those that are as round as apples, or like pears, to those that are as long as cucumbers. In colour they may be green, dark purple or black. In some hot climates people are lucky enough to have them weighing up to 4 pounds each, but those from ¼ pound to ½ pound are more usually seen in our greengrocers.

As a nourishing, satisfying, low calorific food for slimmers, they are unbeatable. They contain vegetable protein oil, eleven different vitamins and seventeen micro-nutrients in the form of trace elements. Babies in Israel are said to be weaned on them because of their easily digested, concentratedly nutritious pulp.

They are delicious with salads or as a meal by themselves. They can be eaten with natural yogurt or cottage cheese or anything of a bland, mild flavour

because they have the consistency of butter without its
weight-giving propensities, and the flavour of fine fresh
nuts.

Strong dressings only spoil their delicate flavour, and
the best non-fattening filling, after their stone has been
removed, is lemon juice with a drop or two of olive oil.

It is amusing to read that the Victorians had the right
idea:

> 'Avocado Pears', often called 'Alligators' are much
> esteemed by natives of the West Indies, also by their
> cats, dogs and horses, because they have a rich
> flavour, which gains upon the palate. Indeed those
> who have no palate foolishly make use of some spice,
> sugar or pungent substance like wine to give them a
> false poignancy.

If when you have a ripe one ready to eat you can bear
to save half, you will find that it makes a soothing,
nourishing face-mask if its pulp is mashed with a little
honey and cream.

Although they look so untypical, the scientific
description of Avocado Pears is that they are berries.
Their stone is their solitary seed. This germinates easily
indoors in a potful of damp soil and soon grows into a
shiny-leaved house plant.

Avocado Pear
(Persea gratissima)

3
BLADDERWRACK
(Fucus vesiculosus)

'Great tufts of bladderwrack spring from the lower stones, and now lie flaccid about, awaiting the returning tide to erect them and wave their leathery leaves to and fro.' So wrote Philip Henry Gosse, the naturalist, in 1877 in his book *A Year at the Shore*. He possibly knew its virtues as a herb, but in comparison with plants that grow on the land, very few seaweeds have become traditionally used for culinary or medicinal purposes.

Bladderwrack, a common brown seaweed which is found all round the coast of Britain, has had the reputation of being an 'anti-fat' herb for centuries. It is also called Sea-kelp, Black Tang, Seawrack and Bladder-weed, which must have come from the poppable air-bladders on its fronds.

Georgian women herbalists used it for 'the reduction of adipose tissues and corpulency', and they made their potions by boiling down the whole plant into a concentrated essence which they administered drop by drop in a wineglassful of water, 'three times daily'.

Modern qualified medical herbalists sometimes prescribe Bladderwrack pills for those who want to lose weight. It is important to consult one of them or a doctor before trying this herb, for it is now known that it acts on the ductless glands, especially on the thyroid.

As a sea herb, Bladderwrack is rich in iodine, iron and calcium, and as well as being a weight reducer it can have a beneficial effect on faulty finger- and toe-nails and also on the hair.

Farmers used to collect it after storms, when it had

been cast ashore, and use it in the production of ground-fertilizing 'kelp'. It used to be cultivated for that purpose on certain stretches of rocky coast and its gathering provided work for the women and children. In the Channel Islands it was used as a plain manure for the fields and is possibly still used there now as it is in Britain by some natural compost-using growers.

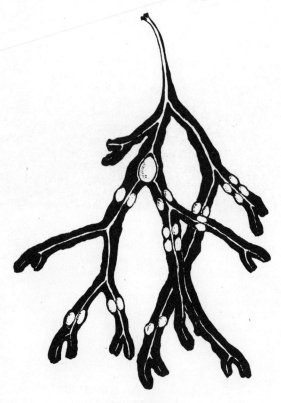

Bladderwrack
(*Fucus vesiculosus*)

4
BLUE FLAG
(Iris versicolor)

This Iris varies a great deal in the colour of its flower and is really the purple, slate blue or even dull red counterpart of the wild British Yellow Flag. The Blue Flag grows in North American swamps and boggy places and can easily be grown in rich damp conditions in this country. It has indeed been known here for nearly two hundred years and was named in Europe, in 1753, by Linnaeus the famed Swedish botanist.

Its home name in America varies according to district, from 'Liver-lily', 'Dagger-flower', and 'Snake-lily' to 'Swamp-lily', and it has been used as a medicinal herb by the North American Indians, from whom we have had many other curative herbs, for centuries.

Herbalists now suggest it as a remedy for sluggish metabolism and a liverish condition. It actually increases the flow of bile and affects the glands and gastrointestinal mucous membranes. It must only be taken on prescription, but it will speed up the working of the digestive and other systems and thereby helps those who are trying to lose some unnecessary weight. It is a general tonic, helping to stimulate the body and assist secretions and excretions.

It also has a reputation for helping women in childbirth: 'Iris helpeth a woman to speedy deliverance and maketh rainbows to appear.'

It is the root of this Iris that is used curatively and it is said to taste acrid and generally unpleasant. Many other Irises have also got a reputation for being useful. The dried root of our Yellow Flag was once used as a snuff, very painfully, 'to cleare the head of mucous', as well as in an 'oyntement' to rub on the sores or galls of horses.

The native Stinking Iris has been credited with cramp-curing potential, and the tubers of the Florentine Iris, or 'Fleur de Lis' (or 'de Louis' after one of the numerous French kings of that name) are the source of Orris powder which used to be used in all the most expensive perfumes. It smells of violets and was ground and made into face powder, too.

Plutarch said that the name 'Iris' meant 'eye' in 'the ancient Egyptian tongue'.

Blue Flag
(Iris versicolor)

5
BORAGE
(Borago officinalis)

Borage has been planted in herb gardens for centuries. It has cobalt-blue, starry flowers, much loved by bees, and once it has been planted it will spread without any help from a gardener.

'I, Borage, bring courage,' is a phrase that is attributed to Pliny and one that may well have been uttered without any thought of the meaning behind it, until scientific herbalists realized that this plant does indeed stimulate the adrenal glands, especially if they are sluggish, and thus really proves helpful both mentally and physically. So this is the way in which Borage can help slimmers.

It is interesting that its old reputation has now been proved, for Gerard, the seventeenth-century herbalist, said: 'Those in our time do use the floure in sallads, to exhilerate and make the minde glad.'

Both flowers and young leaves are good in salads, the former in a decorative sense and the leaves to give a faint cucumberish flavour. The blue flowers are famous as decorations in Pimm's No. 1, or in claret or fruit cups. They even improve the look of homemade lemonade. They have, always been said to be cooling, long before the use of ice in drinks, and one of their country names is 'Cool Tankard'.

In Tudor and Stuart tapestry and other needlework, the bright blue flowers can often be seen. Embroideresses were thought to have picked and arranged the flowers in patterns, which they then copied. Perhaps the Borage cheered their melancholy, for this power was also credited to it:

> Borage and Hellebore fill two scenes
> Sovereign plants to purge the veins
> Of melancholy and cheer the heart
> Of those black fumes which make it smart.

(Burton, from his famous *Anatomy of Melancholy*)

Borage tea, made from an infusion of young leaves either fresh or dried, is rich in potassium and makes a stimulating, tonic drink, especially if a few blue Borage flowers are floated on it.

Borage is an annual and the young leaves when they first start growing are very rough, so that they qualify it for another country name of 'Ox-tongue'.

Borage
(Borago officinalis)

6
CELERY
(Apium graveolens)

Celery, or Smallage, grows wild near the sea. It is a tough, fibrous, aromatic plant with deep green leaves which is left ungrazed by cattle. It has been cultivated and blanched and used as a vegetable since ancient Egyptian times.

The Victorians in their pseudo-refinement objected to the noise that Celery eaters made. A drawing from an autograph album of a family at table shows a young Victorian husband with a stick of Celery which has provoked the verse below:

> Hush, hush, my dear,
> Let me not hear
> You munch your food
> So near my ear.

Curiously enough, their gardening correspondents extolled it, saying that 'when carefully grown, it becomes a sweet, crisp and juicy and most agreeably flavoured asset to the diet'.

Its use now, chopped as a salad ingredient or by itself eaten raw with cottage cheese, roast meat, grilled steak, or even fish, is both slimming and beneficial in other ways. A few Celery stems chewed instead of sweets when the urge for them is strong, will not add any weight, and as Celery juice is a tonic to the system and said to be a solvent of uric acid, it is helpful to anyone with any rheumatic tendencies. It is also a digestive and tends to provoke, according to one of the old herbalists, 'a goodly flow of urine'. In other words, it is a mild diuretic.

Big heads of tender, near-white leaf stems with crisp 'hearts', are difficult and time-consuming to grow. Amateur gardeners need much space and effort to produce them, but a variety called Celeriac is far simpler as it needs no trenching. It has the same appetizing flavour (and useful properties), but does not look as appetizing with its swollen base. This can be grated for salads, or boiled as a tender cooked vegetable.

There used to be an idea that the green parts of cultivated Celery were poisonous. Now that the pale green, real Celery heads are imported so freely from Israel, it appears perhaps that all the terrific effort of complete blanching may have been more to improve their looks and flavour. The imported pale green heads are not as strong in flavour, to some tastes, as home-grown white Celery.

Celery
(Apium graveolens)

7

CHICORY
(Cichorum intybus)

Chicory, or Succory, is a tall wild perennial plant with big Wedgwood-blue daisy-like flowers, each bigger than an old half-crown but lasting only for one day.

Cultivated Chicory, a variety called *foliosum*, has been developed from our wild plant. Chicory has been used as a vegetable for generations and has been forced and blanched by digging up second-year roots in the winter and potting them up three or four together in a big flowerpot. This is kept in a dark, warm boilerhouse until the fresh fat white shoots appear.

In Victorian times the French used it in winter salads and called it 'Barbe du Capucine', making sure that the 'young leaves were thoroughly etiolated'.

Only a few gardeners in this country still cultivate it like this, but commercial growers in Belgium and France produce all that we see in the shops. Greengrocers keep it in the boxes in which the growers have packed it, covered with dark blue tissue paper to prevent the light turning the tips of the young leaves green and thus making them bitter.

As a salad plant, Chicory is slimming and contains valuable traces of mineral salts. It is delicious with slices of orange or grapefruit.

Endive (*Cichorum endivis*), another species, is also a useful vegetable for winter salads. It can be used green or blanched. It too is cultivated in quantity on the Continent, in Holland and France.

Chicory roots are also used. Tea made from them after they have been dried is a good general tonic and improver of slowed-up metabolisms. It is also said to be good for rheumatism and gout. It is made from the first-year roots and has been used, curatively and as a beverage, as well as an admissible additive to coffee for over 250 years.

An English botanist, in the middle of last century, writes of

'Chicorée à café' which is cultivated extensively in France for the sake of its roots, which are taken up in the winter season, cut into squares, dried artificially and afterwards roasted. Then they are ground with their coffee for which they serve as an adulteration. There are, however, those who assert that it is to this admixture of Succory root that the superior flavour of the French to the English coffee is attributed.

Chicory is now used here in a mixture with many ready-prepared ground coffees.

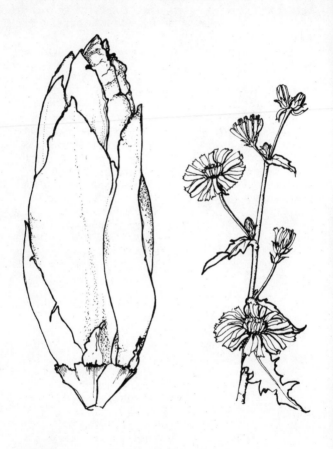

Chicory
(Cichorum intybus)

8
CLEAVERS
(Galium aparine)

Cleavers is one of the favourite slimming herbs, and has been for centuries. 'It is familiarly taken in broth to keep them lean and lank that are apt to grow fat', wrote Culpeper in the middle of the seventeenth century.

It is a very common British weed and has endless country names which vary from 'Goosegrass', 'Goosebill' (all poultry seek it out to eat, especially in the spring) to 'Hayriffe', 'Burrweed' and 'Little Sweethearts'. The Greeks called it 'Philanthropon' or 'Man Lover' because its small curved hooks on the stems, leaves and fruits, catch and cling to anyone who touches it. It can be a tiresome weed in some places, with its clinging, climbing habit.

Cleavers belongs to the same botanical family as the Bedstraws which were used as litter in houses for strewing the floors, long before carpets were readily available or cheap enough for everyone to have. Coffee, Quinine and Ipecacuanha all belong to this same family.

As a medicinal herb, Cleavers tea makes a good tonic and blood purifier, as well as a slimming drink. It has always been thought to be good for the complexion. The tea can be made by putting a handful of the fresh or dried herb, which is available from health stores, in a jug and pouring half a pint of boiling water on it. It should steep for fifteen minutes and then be drained and drunk, a wineglassful at a time, three times a day. This tea makes a soothing sunburn lotion.

Cleavers contains three vital mineral salts, calcium, sodium and silicon. The plant's flowers are very small and white and the petals are arranged in the shape of a Maltese cross. The twin, spherical fruits are hooked and can be dried, roasted and ground to make a harmless substitute for coffee; they have a faintly salty flavour.

Cleavers
(Galium aparine)

9
DANDELION
(Taraxacum officinale)

Dandelions are one of our most familiar wayside flowers, 'Fringing the dusty road with harmless gold', as Lowell, the American author, said. In the old days Dandelion flowers used to be gathered for wine-making and now, again, these sun-heads are often picked to be converted into homemade wine.

The plant has many uses. A few young leaves in a salad, together with Lettuce and Sorrel, make a new flavour: 'With this homely sallet Hecate entertained Theseus', according to the diarist John Evelyn.

The leaves are supposed to counteract biliousness and to be beneficial for anyone who has an excess of uric acid or for sufferers from rheumatism and arthritis.

Unfortunately Dandelion leaves, unless very young, are very bitter and tough. This was one of the bitter herbs used in the Feast of the Passover. They taste better if they are blanched, and some enthusiasts lift a root up, in the autumn, plant it in a pot like the Chicory growers and keep it in a warm, dark cellar. They pick and eat the young etiolated leaves as they grow.

Dandelions are rich in mineral salts and have calcium, iron, magnesium, potassium and silicon in their make-up. They were used as a cooked herb by settlers in the U.S.A. and Culpeper mentions them as pot herbs, boiled 'with a few Alisanders' which are now known as Alexanders, an aromatic, tall and beautiful wild plant which grows near the sea.

Dandelion roots are roasted for coffee and Dandelion coffee is a much more slimming and healthy drink than the ordinary coffee. A Victorian doctor recommended it to his patients: 'When drunk at night it produces a tendency to sleep, instead of exciting wakefulness and it may be safely used as a cheap, wholesome substitute for the Arabian berry.'

But let old Culpeper have the last word. He was, apparently, suffering when he wrote this from a certain amount of derisory comment from the doctors of his time:

This common herb is full of virtue and that's the reason the French and Dutch so often eat them in the Spring; and now if you look a little further you may see plainly without a pair of spectacles, that foreign physicians are not so selfish as ours are, but more communicative of the virtues of plants to their people.

Dandelion
(Taraxacum officinale)

FENNEL
(Foeniculum vulgare)

Fennel seeds, which are delicious and taste faintly of ani-seed, or those of Angelica, Dill, Parsley and even a few Sunflower fruits, can be chewed instead of sweets. They all contain non-fattening nutritive elements, but the Sunflower's of course are oily and not as slimming as the others.

Fennel seeds seemed to have earned the name in the early eighteenth century of 'go-to-meeting-seeds'. They were chewed, unobtrusively, during long church and chapel services. Earlier than this they also used to be used to plug key-holes, 'to keep the ghosties away'.

Fennel has been known as a digestive and as a flavouring herb since the time of the ancient Egyptians. It had also long been used as a cure for bad breath, an old herbalist's advice being to 'Chew a leaf of Fennel every day at dawn or else Orris [the root of the Florentine Iris], or boil Parsley to drink, or crush Cumin seed.'

Fennel also was reputed to be an eye strengthener: 'A serpent doth so hate the ash tree that she will not come nigh the shadow of it, but she delights in Fennel very much which she eates to cleare her eyes.'

It was also put in 'Gripe Water' and given to babies to soothe them, but it had plenty of sugar with it. Its flavour alone is delicious and this is how it can be used as a slimming herb, either as a tea or, as the old botanist William Coles said, by putting the 'leaves, seeds and fruit of our Garden Fennel into broths for those that are fat, to abate their unwieldiness'.

There are several forms of cultivated Fennel for the

garden and all are tall, elegant plants with very finely divided leaves and umbels of minute yellow flowers.

Finnochio, or Florence Fennel (see drawing on page 38) can be used as a delicious vegetable, for its base swells rather like a softer Celeriac, and then it should be 'earthed up and eaten three weeks later'.

As well as being grown in the old days as a medicinal and culinary herb, Fennel was used for freshening bee hives (which, of course, were then straw skeps). This comes from a seventeenth-century herbal:

> You shall perfume the Hive with Juniper, and rub it all within with Fennel, Hyssop and Time-flowers, and also the stone upon which the Hive shall stand . . . for in all clenlinesse and sweetnesse the Bees are much delighted.

Fennel
(Foeniculum vulgare)

11
GRAPEFRUIT
(Citrus paradisi)

At the moment Grapefruit seems to be regarded as the most slimming fruit of all. Even a writer in the *British Medical Journal* has seen fit to say that Grapefruit juice has 'the physiological action of stimulating the appetite and provoking salivary and gastric digestion'. Half a Grapefruit before each meal means that the food that follows it is well and easily assimilated and thus does not get at all hung up in the process of use or elimination.

It is strange to realize that Grapefruit has only been popular in this country for about forty years. It grew wild, apparently being a known variety of Citrus fruit that was not thought to be worthwhile except in a few places.

In China, years ago, its virtues were recognized and it was called 'Huoss-gien' or 'Sweet Ball', and as well as eating it, the women realized its value as a tonic in pregnancy. They also made its flowers into fragrant cosmetics.

Grapefruit has to be grown in hot countries for the production of the finest fruit. Since it was first cultivated, different strains have been selected by judicious plant-breeders, so that the thinness of the skins and the juiciness of the pulp have both been vastly improved.

In Japan, during last century, the fruit was said to grow 'as big as babies' heads'. The pulp was described as being red, white or yellow in colour.

A Victorian botanist refers to it by the unfamiliar name of 'Shaddock'. It seems that it was a Captain Shaddock, a naval man, who had tasted it, and liked it, in the East Indies and had taken some young trees to the West Indies, where it was henceforth grown. But it has had other names as well including 'Forbidden Fruit', 'Pomelo' and 'Pampelmouse'.

The name 'Grapefruit' appears to have come into common usage from the fact that the 'many segmented berries, like big pale yellow oranges, with their bitter-sweet pulp, grow in clusters, rather in the manner of grapes'. But also, of course, some of the best fruit can have the flavour of fine grapes, which may too have accounted for the name by which we now know it.

Grapefruit
(Citrus paradisi)

JUNIPER
(Juniperus communis)

Juniper is one of the three native coniferous trees in Britain and it can hardly even be called a tree, for on the chalk hills where it still grows in unploughed areas, it reaches only shrub height and status. Its needle-like spiny leaves grow in whorls of three and its insignificant female flowers develop over a period of two years into currant-sized, blue-black, berry-like fruits.

It is these fruits that are used medicinally, to form a diuretic or kidney stimulator. *This herb should never be tried as a slimmer, unless it is prescribed by a qualified medical herbalist.* In some cases of over-fatness it can be useful and most efficacious, but in others it can upset the general balance of the whole system.

'Genièvre' is the French word for Juniper, and 'Genièvre Water' or 'Juniper Water' used to be called gin. Indeed Juniper fruits were long used to flavour gin until, it seems, in the nineteenth century when a Professor remarked that it was 'Wholly unconscious of their presence'.

Dr Coffin, at a later date, suggested soaking Juniper berries in gin and throwing out the gin and eating these berries. It was he who suggested that they should be eaten as a protection against stomach upsets when travelling abroad.

The tips of Juniper branches used to be burnt in sickrooms and hospitals 'to render the air wholesome' and also in kitchens to get rid of cooking smells.

Juniper has been considered as a magic shrub for

many centuries. A sprig of the plant carried on the person was thought to be a protection against devils, evil spirits and wild animals. It was also used against thunderbolts. Stables in Italy, instead of our customary horseshoe, had Juniper branches nailed over the door to protect the horses within.

The Juniper tree may have earned its tradition of protection from the legend in which it sometimes joins Rosemary, or even replaces it, as being the tree which sheltered and saved the life of the Virgin Mary when she was fleeing from Herod.

Within living memory, it used to be said by old huntsmen in Sussex that if hares or foxes sheltered in the lee of a Juniper bush their scent was lost to hounds.

Juniper
(Juniperus communis)

MATÉ TEA
(Ilex paraquayensis)

Modern herbalists say that Maté Tea is one of the safest and easiest herbs to help slimmers. Its chief virtue, for this purpose, is that it is so satisfying that it enables arduous work to be done on less food.

It is also very good for rheumatism and is a mild aperient and a tonic for the kidneys. It is soothing to those who sleep badly and also a great help in increasing intellectual lucidity. Indeed they make it sound a veritable panacea for all ills.

Maté Tea is made from the leaves of a tree that originally grew in Paraguay in South America. It is now cultivated in other tea-growing areas. The tree is related to our Holly, although its leaves (see drawing on page 45), are nothing like our usual spiny-edged Holly leaves.

The natives used to brew up the tea by putting a handful of leaves in a special teapot called a Maté and then imbibe the hot liquid through the spout of it. The spout was called the *bombilla* and was perforated 'with holes at the top to prevent the swallowing of the pulverised herb'. The Maté was passed round from one person to another and the vessel filled up with hot water as fast as possible. It is easy to imagine the distaste, especially in the Victorian era, for this community drinking, and a contemporary account mentions that 'the Europeans present, not fancying this habit, served theirs in little glass tubes'.

In the last century there seem to have been three grades of Maté Tea. *Caa-cuys* was made from the young leaves, still in bud; *Caa-mini* from expanded leaves stripped off their midribs, and *Caa-quazu* in which the whole leaves were roasted without preparation. Unfortunately there appears to be no surviving information as to their differing merits.

Ordinary Maté Tea drinkers can sweeten their drink with a little honey if necessary, but many people prefer it plain or with a squeeze of lemon juice. The infusion can be renewed quickly by the addition of more boiling water, but it must be drunk fresh or it will go black.

In Paraguay it is called *Yerva-maté*, and in France *Herbe du Paraquay*.

The flavour is impossible to describe. There is a faint smokiness and herbalists liken it to tea made from Mallow leaves. The Maté leaves are passed through heavy smoke before being dried in hot sunshine and then artifically heated and ground.

Maté Tea
(Ilex paraquayensis)

14
MUSHROOM
(Psalliota campestris)

Mushrooms are amongst the most delectable of the fruits of the earth that are actually good for slimming. They are low in calorific value but contain vitamins and traces of mineral salts that are essential to good health and they have such a delicious flavour.

But slimmers must not cook them in fat. Clean, cultivated button mushrooms can be sliced and eaten raw as long as they are very fresh. The Mushroom Consumer Service say that they do not even need washing, merely wiping with a damp cloth or rinsing in a colander. This organization has useful free leaflets of Mushroom recipes which can be obtained by sending a 2½p stamp to their Public Relations Officer, at Agriculture House, Knightsbridge, London SW1. If anything can add variety to slimming menus, Mushrooms most certainly can and some of the recipes sound mouth-wateringly appetizing.

Mushrooms have been eaten for hundreds of years and picked with the greatest appreciation from the fields; they have also been viewed with suspicion. This description comes from the *Great Herball* of 1526, and seems to mark their debut in print in this country:

Fungi ben muscherons . . . there be two manners of them, one moner is deadly and sleeth them that eateth of them and be called tode stoles and the other doeth not. They that be not deadly have a grosse, glemy moystur, that is disobedient to nature and digestion and be perillous and dredfull to eate, and therefore it is good to eschue them.

Surely that must have put people off for hundreds of

years? In Ireland, mushrooms are still viewed with suspicion in the 1870s when there were great crops of mushrooms that appeared in the fields which were picked and shipped to English markets. The story was that they didn't fancy eating them themselves, as the cows would not touch them.

There are plenty of other flavoursome edible fungi which are viewed by the majority of people with suspicion still. A few enthusiasts seek morels, the elusive truffles, beef steak fungus, puffballs, edible boletus, and chanterelles, which they cook and eat appreciatively. *It is very important indeed to know all fungi well before experimenting with these obscure delights, for a few are dangerously poisonous.*

Field mushrooms are scarce now but can still be found where any of their fields are unploughed and unsprayed. Local pickers get up at dawn to find them during the months from June to October. It is interesting to notice, apart from the delight of finding them at all, how they grow in ever-widening circles, or 'fairy-rings', which used to arouse local superstitions, not so much about fairies, as about witches, the Devil, or even dragons.

Cultivated mushrooms are not new, for the collecting of spawn started in France at least 300 years ago. It was in the U.S.A., early in this century, that pure mushroom spawn was produced by cultivating living tissue from a young mushroom. Then came spawn grown from mushroom spores and these discoveries were the beginning of our great new industry here.

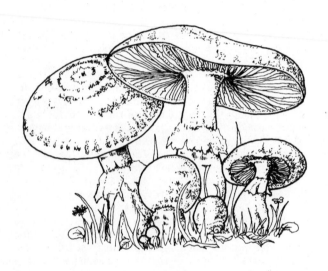

Mushroom
(Psalliota campestris)

15

PARSLEY AND OTHER RAW FLAVOURERS

On the whole, the herbs that can be chopped up raw or used in juice form in vegetable cocktails can usually be used cooked as well. The flavour and the vitamin content are higher when they are raw and less of them is needed to give an unusual tang to such ordinary dishes as omelets, savoury rice or scrambled eggs.

Garlic is so strong that a mere rub round the salad bowl is enough for many people. It is a great provider·of mineral salts, but must be avoided, like Chives and Onions, by those who have any form of sulphur allergy. Garlic has been credited with magic and antiseptic powers since the days of early civilizations. It is said to have kept the pyramid-builders, in ancient Egypt, strong and healthy. The Ancient Greeks would not allow people who ate it into the temples of Cybele.

Rue also had magic powers assigned to it and a very small amount of the leaf chopped up in salads adds a piquancy that is different from that given by the equally bitter dandelion leaves. It gets its pleasing name of 'Herb of Grace' from the tradition that a twig of it used to be used to sprinkle holy water. Country people still use raw, crushed Rue leaves to apply to bee stings.

Everyone can grow Chives or 'Rushleeks' in a garden or even in a pot on a windowsill. Parsley, though one of the most useful of all the slimming herbs, is more difficult. It is all right once the seed germinates and 'gets away', but it seems to be fussy about its situation and it used to be said that when it grew well, it was a sign that the woman of the house wore the trousers! The seed was also thought to go down to hell and back again before it started to grow, because the seedling took so long to appear.

It is well worth any effort to get it to grow for its delicious fresh flavour and its high vitamin and mineral content as well as for its marked tonic and body--stimulating effect.

Some people delight in the addition of sour Sorrel to salads, the leaves of which should be finely chopped. Mint leaves, chopped raw, are a vast improvement to many cooked vegetables and other dishes.

A little experimentation with harmless flavouring herbs can often result in creating a unique dish. One Rosemary leaf, chopped into a portion of cottage cheese, improves it greatly. Grated root of Horse-radish, used with discretion, improves beef whether it be from a joint, or a mere lone steak.

The young leaves of Costmary, Sweet Basil, French Tarragon, Lovage, Sage and Sweet Cicely may all be added to salads and the last is useful to slimmers in a different way, too. For a few Sweet Cicely leaves added to tart fruit like Gooseberries as they are being cooked, reduces by nearly half the amount of sugar needed to make them edible.

Parsley and other raw flavourers
Bay, Garlic, Rue, Parsley,
Chives, Mint

16
PEPPERS
(Capsicum annuum)

These Peppers grow to many sizes, shapes and colours and are generally known as Sweet Peppers. They can all be used raw, in salads, or cooked, and stuffed.

Another species, the smallest of all, are the bitingly hot 'Chillies' or 'Bird-Peppers' which are useful for curries and about which our forefathers raved. Many reminiscences have flowed about them, and among the favourites is the one about the man who was dining at Buckingham Palace with Queen Victoria. He bit on a Chili and took it out of his mouth in his fingers. He is reputed to have been so disturbed in case the Queen had seen him, that he justified himself by saying 'only a damned fool would have swallowed that', which cast him even deeper into the mire with the prim Queen. It has been suggested that at one stage, these Chillies were thought to postpone, or even cure, delirium tremens.

Cayenne pepper is made from the dried and powdered Chillies. Paprika comes from larger Capsicum. Both are rich in vitamin C and are cultivated in the tropics and in Europe, in Hungary and the Mediterranean countries.

Peppers can be grown indoors as ornamental house plants. They are not difficult to raise. They make extremely decorative plants when they come into fruit and are hung with their green and red peppers. The small, round-fruited Christmas Cherry is another kind of Capsicum and they are all in the same botanical family as potatoes and tomatoes.

Perhaps the inventor of this Victorian recipe, which

still sounds both practical and attractive, had heavy
drinkers in mind.

Make a mixture of sliced cucumber, shallots and
onions cut very small, with a dash of lime juice and
Madeira wine. Add a few pods of 'bird-peppers'
[Chillies] well-washed and you will seldom fail to
provoke the most languid appetite.

Strangely enough, minute homoeopathic doses of
Capsicum cover a wide panorama of symptoms that they
will cure. These symptoms range from acute earache,
especially from the inflammation of mastoid; to bleeding
piles, sore throats and homesickness.

Peppers
(Capsicum annuum)

17

ROSE GERANIUM AND OTHER COOKING FLAVOURERS

The value of flavouring herbs for clear soups, casseroles, savouries and with cooked vegetables cannot be over-stressed for those who are on a slimming crusade. No one denies that the range of non-fattening foods is small in comparison with the great variety that non-slimmers indulge in, and therefore one of the weight-watcher's problems is to avoid getting bored with their diet.

There are a great many culinary herbs that give different and interesting tastes. Mint (see page 50), perhaps, is the most commonly used (and it can certainly be used raw as well). Mint tea is a very old remedy for indigestion.

Anyone can grow but not always control one of the better flavoured varieties of Mint in a garden, and it is easy to dry for the winter, by picking it green and hanging it upside down in a warm shady place in the summer.

Marjoram; Dill, with its name from the Norse *dilla* which meant 'to lull', is delicious with cucumber; Sweet Cicely; Bay leaves; Caraway, with its country reputation for never getting lost, so that people used to give it to their fowls and pigeons; and Thyme, are all excellent flavourers. Thyme was thought to have magic properties and a sixteenth-century recipe for it was 'to enable one to see the fairies'.

Cumin seed, used sparingly as a spice (as well as for breath cleaning, see page 36); Cloves which are the buds from a tall tree growing only in the Moluccas and which in its Latin name of *Eugenia aromatica* com-memorates the name of Prince Eugene of Savoy who had a keen interest in botany and was one of the first known plant protectors; Rose Geranium leaves; the famous Sage, which as well as being such a good culinary herb, makes a fine gargle if infused, for a sore throat; Hyssop, with its 'flowers of heavenly blue'; and Chervil; Lovage; and even a solitary leaf or two of Rosemary inserted into scores on a leg of roasting lamb, should all be tried.

Sorrel is extremely acid; Rue, though graceful in the garden and 'rank-smelling' according to Spenser, is pungent as well as slightly bitter; both can take their places in the experimental flavouring procession.

There are more, of course: Lemon Balm, Southernwood, Winter Savory and Tarragon, the list seems endless, for so far there's been no mention of Tansy or Costmary:

> Anything green that grew out of the mould
> Was an excellent herb to our fathers of old

as Kipling so rightly said.

Cooking Flavourers
Rose Geranium, Rosemary:
Thyme, Sage, Dill
and Cloves

18

SUNFLOWER
(Helianthus annuus)

Sunflowers are one of the largest annual plants that can be grown in gardens. They often grow up to eight or nine feet tall, all in one year. Their great soup-plate-sized flowering heads are made up from hundreds of dark central and golden edging ray florets and are often slightly nodded as if they are too heavy to hold up.

They are natives of Mexico, brought to Europe by the conquering Spaniards. The species that produce the now popular seeds and oil are being grown commercially in Hungary, Italy and Southern Russia.

The huge flowers look like suns drawn by little children and it is often said that they follow the sun, starting in the morning facing east and finishing in the evening towards the west. In actual fact this does not seem to happen in this country: perhaps the sun does not reach the heat or brightness necessary to effect the adjustments needed.

Technically speaking, each 'seed' in the slightly concave centre of the flowering head becomes a nut. Its outer shell is hard and may be black, near-white, or striped in a variety of widths in both black and white. These sunflower nuts are full of nourishment for humans and for the birds that seek them out and feed on them in the garden. Poultry, pigeons and caged birds also revel in them.

If you are weight-watching, eat only a few at a time. They are less fattening and far better for you than sweets or biscuits and they are rich in phosphorus, but they do contain a good proportion of vegetable oil.

Luckily, a few are quickly satisfying.

Sunflower oil, used in some margarines and for salads, is extracted from the 'seeds' by pressure. The residue of shell and 'seed'-covering is converted into nourishing cattle-cake.

Juice from the Sunflower leaves makes a quick healer for cuts and other small wounds.

Helianthus tuberosus, another species of Sunflower which is a perennial and has far smaller heads of yellow-rayed flowers, produces Jerusalem Artichokes. A small helping of these is less fattening than an equal amount of potato.

Sunflower
(Helianthus annuus)

19

TOMATO
(Lycopersicum esculentum)

Tomatoes, which were brought to Europe from Mexico and Peru in the sixteenth century, again by returning Spaniards who had found them so pleasing to eat there, were at first treated with much suspicion by the Old World, and called 'Wolf's Peach'. Possibly this was because their fruits looked too attractive and too flamboyantly red to eat.

They belong, of course, to a botanical family which does include several poisonous species. Even in this country, the feared Deadly Nightshade, the purple Bittersweet, the evil-smelling Henbane and the strange Thorn-apple are poisonous representatives of *Solanaceae*. The family, however, when considered in a more dispassionate manner, has as many plants that are of benefit to man, for it provides the Egg-plant or Aubergine, the Peppers and Potatoes, as well as Tomatoes, even if they all came originally from overseas.

As people began to realize that Tomatoes were harmless, they seem to have changed their popular name to 'Love Apples', but even so they have only been grown in Britain since the nineteenth century.

There are now many named varieties. They form one of our biggest economically sound commercial crops, and in some parts of southern England acres of ground are covered by the glasshouses in which they are grown.

They can also be grown by amateurs, after the young plants have been raised in a greenhouse, but they need a sheltered position out of doors and no frost and they must have plenty of sunshine to ripen them. It is amazing that one ounce of Tomato seed will yield about 2,000 plants.

For slimmers, especially, they provide a non-fattening, interestingly flavoured and textured food which is good raw in salads, or cooked in savouries, broths or plainly juiced. Tomatoes are rich in vitamins A, B and C. Cooking destroys at least half the vitamins.

Tomatoes
(Lycopersicum esculentum)

20

WATERCRESS
(Nasturtium officinale)

Watercress, or 'Watercurse', as it used to be called, is rich in mineral salts and vitamins as well as green chlorophyll, and it is probably one of the most beneficial of our raw salad plants.

There are people who are nervous of eating it even when it has been cultivated by accredited growers in special shallow-bottomed, stream-fed water beds. Nowadays there is little chance of sewage contamination under such conditions, but there is still the fear of the presence of liverfluke parasites, which are carried by water snails. If, though, you wash the Watercress *thoroughly, in salt and water,* this ought to eliminate any water snails.

If you are at all worried, grow Landcress instead. The seed is available from specialist seedsmen and the plants will grow in any garden without much care or attention. The flavour is almost as good as that of Watercress, but the leaves and stems, after being grown in far drier conditions, do naturally lack the succulence of good Watercress. But Landcress is certainly safe.

If you are still an avid Watercress eater, and some people really yearn for it, NEVER be tempted to pick any plants growing wild in a ditch or stream. For one thing there are other plants with leaves that may look very like those of Watercress, but which are harmful, if not actually deadly poisonous, that grow just as lushly and commonly in damp places and in ponds and running steams. For the second reason, there is the near certainty that such places are now polluted or contaminated.

Although only a few salad plants are mentioned in this book, others, like Cucumber, Mustard and Cress (so easy to grow all through the year) and Corn Salad are all beneficial to health and useful to slimmers. Other raw vegetables, like Carrots, Brussels Sprouts, crisp Cabbage and Cauliflower too, can be delicious when grated or chopped. The curious thing about them is that if they are properly masticated, they are less indigestible than when they are cooked, and certainly a smaller quantity is needed to satisfy the grossest appetite.

Watercress
(Nasturtium officinale)

THERAPEUTIC INDEX

Antiseptic, Garlic, Juniper.
Aperient, (mild), Maté tea.
Bad breath, Fennel.
Bee stings, Rue.
Biliousness, Dandelion.
Blood purifier, Cleavers.
Cuts, Sunflower.
Depression, Borage.
Digestive, Angelica, Blue Flag, Celery, Fennel, Grape-
fruit, Mint.
Diuretic, (mild), Celery, Dandelion, Maté tea.
Gargle, Sage.
Gout, Chicory.
Liverish condition, Blue Flag, Dandelion.
Rheumatism, Celery, Chicory, Dandelion, Maté tea.
Sleeplessness, Dandelion, Maté tea.
Sore throat, Sage.
Sunburn, Cleavers.
Tonic, Angelica, Blue Flag, Borage, Celery, Chicory,
Cleavers, Watercress.
Uric acid, Celery, Dandelion.